21st Century Junior Library

Grasshopper

by Susan H. Gray

CHERRY LAKE PUBLISHING * ANN ARBOR, MICHIGAN

Published in the United States of America by Cherry Lake Publishing
Ann Arbor, Michigan
www.cherrylakepublishing.com

Content Adviser: The Entomological Foundation (www.entfdn.org)

Reading Adviser: Marla Conn, ReadAbility, Inc

Photo Credits: © Christian Musat/Shutterstock Images, cover; © Leena Robinson/Shutterstock Images, 4; © Paul Reeves Photography/Shutterstock Images, 6; © Brian Maudsley/Shutterstock Images, 8; © wonderisland/Shutterstock Images, 10; © suradech sribuanoy/Shutterstock Images, 12; © Kirill Ilchenko/Shutterstock Images, 14; © Dr. Morley Read/Shutterstock Images, 16; © Katarina Christenson/Shutterstock Images, 18; © Tyler Fox/Shutterstock Images, 20

LIBRARY OF CONGRESS CATALOGING-IN-PUBLICATION DATA

Gray, Susan Heinrichs, author.
Grasshopper / Susan H. Gray.
pages cm.—(Creepy crawly critters)
Includes index.
ISBN 978-1-63362-590-7 (hardcover)—ISBN 978-1-63362-770-3 (pdf)—
ISBN 978-1-63362-680-5 (pbk.)—ISBN 978-1-63362-860-1 (ebook)
1. Grasshoppers—Juvenile literature. I. Title. II. Series: Creepy crawly critters.

QL508.A2G744 2015
595.7'26—dc23 2014048649

Cherry Lake Publishing would like to acknowledge the work of the Partnership for 21st Century Skills. Please visit www.p21.org for more information.

Printed in the United States of America
Corporate Graphics

CONTENTS

Grasshoppers have very long legs.

Amazing!

Have you ever seen a grasshopper up close? It has two **antennae**, or feelers. Its eyes are enormous. Wings lie flat against its back. And its **hind** legs are huge! Grasshoppers are amazing insects.

Grasshoppers eat leaves and grass.

Hoppy Days

What do grasshoppers do all day? They sit in the sun, hop around, and eat. Grasshoppers mostly eat plants. They gobble up leaves and grass. They chew on flowers and seeds. Sometimes, they eat dead bugs. Then they hop off to find **shelter** or more food.

Adult grasshoppers use their wings to fly.

Grasshoppers sure know how to get around. They use their six legs for walking. But walking is slow. To move quickly, they hop or fly.

Baby grasshoppers are called **nymphs**. The nymphs cannot fly. They have no wings at first. But soon, they grow little wing pads. Four wings will grow out of the pads by the time the insect is a grown-up. Adult grasshoppers use all four wings when flying.

A grasshopper's powerful legs help it jump.

To jump, grasshoppers use their big hind legs. First, they bend their legs up tight. They lower their bodies closer to the ground. They **pause** for just a second. Then they straighten their legs and spring into the air.

Make a Guess!

A grasshopper's hind leg has two main muscles. It uses one muscle to bend its leg before jumping. What does it use the other muscle for?

This grasshopper is laying her eggs in the dirt.

Hatching, Eating, and Talking

Mother grasshoppers lay eggs. First, they dig little holes in the dirt. Then they lay their eggs inside. Usually, eggs hatch in the spring. The new babies start eating right away. At this time of year, fresh young plants are **budding**. The babies have plenty to eat.

These locusts are swarming a field of crops.

Grasshoppers stay wherever there's plenty of food. They live in fields, backyards, and grassy lots. They also live where farmers grow **crops**.

Some grasshoppers travel in **swarms**. These grasshoppers are called **locusts**. Millions of locusts can swoop in on a crop. They can destroy it by eating everything in sight.

This grasshopper pair found each other by making chirping noises.

Sometimes, grasshoppers talk to each other. They talk by making loud sounds with their bodies. To do this, they scrub their legs against their wings. Or they move their wings against each other.

Think!

Grasshoppers have ears on their sides. They can hear the sounds made by other grasshoppers. Why is it important for grasshoppers to hear one another? What sorts of things are grasshoppers telling one another?

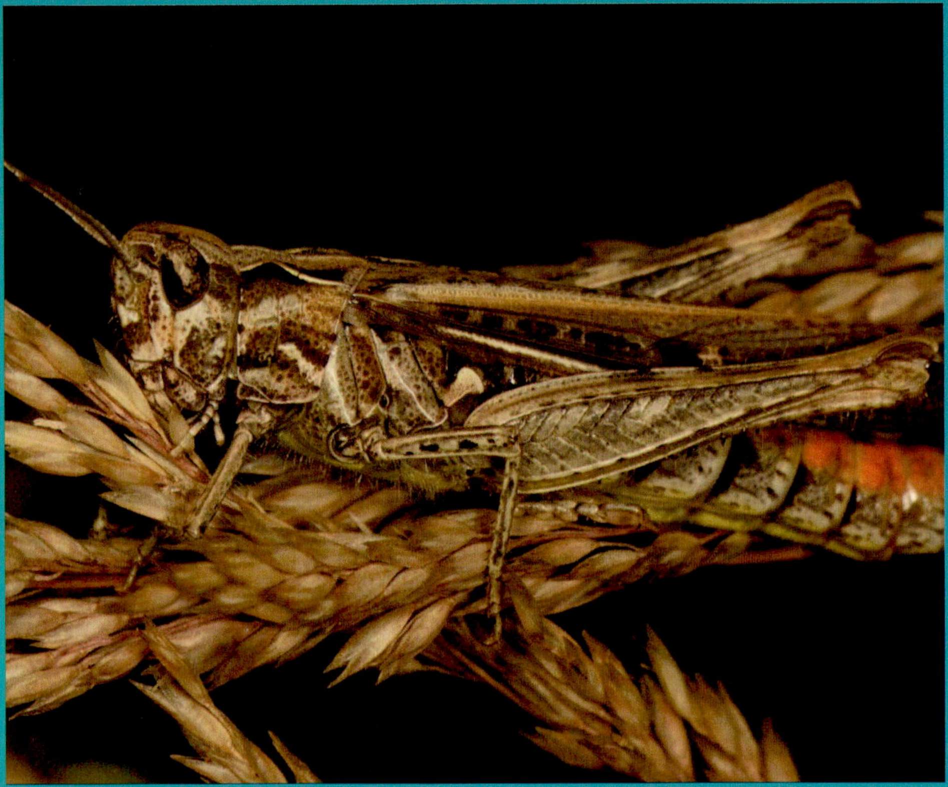

This brown grasshopper blends in with the wheat.

Yuck!

Grasshoppers love to eat plants. But there are **predators** that love to eat *them*. Frogs, lizards, snakes, mice, birds, and large spiders eat grasshoppers. When grasshoppers get caught, they often spit

Look!

Grasshoppers have ways to escape danger. This grasshopper hides right out in the open. Predators cannot see it. What makes this grasshopper so hard to find?

Lubber grasshoppers like this one are common in the southeastern United States.

nasty, brown juice. The surprised predator then drops its **prey**.

Grasshoppers can do lots of things. They jump, and they fly. They dig holes for their eggs. They make noise with their legs and wings. They hear through their sides. And they spit when something scares them. We have so much to learn about these incredible insects!

GLOSSARY

antennae (an-TEN-ee) thin sensory organs on the heads of insects

budding (BUD-eng) just starting to grow

crops (KROPS) plants grown for food or other uses

hind (HYND) back or in back

locusts (LOW-kusts) types of grasshoppers that travel in swarms

nymphs (NIMFS) a name used for some insects, such as grasshoppers, that have not yet become adults

pause (PAWZ) to wait for a short time

predators (PRED-uh-turz) animals that hunt, kill, and eat other animals

prey (PRAY) an animal that is hunted by other animals for food

shelter (SHEL-tur) a safe place

swarms (SWARMZ) very large numbers of animals that travel together

FIND OUT MORE

BOOKS

Ashley, Susan. *Incredible Grasshoppers*. New York: Gareth Stevens Publishing, 2012.

Bodden, Valerie. *Creepy Creatures: Grasshoppers*. Mankato, MN: Creative Paperbacks, 2014.

WEB SITES

BioKids—Kids' Inquiry of Diverse Species: Grasshoppers

www.biokids.umich.edu/critters/Acrididae/
This site has information on different grasshoppers and how they eat, grow, and behave.

Bug Facts: Grasshopper

www.bugfacts.net/grasshopper.php
This site has good, basic information on where to find grasshoppers and how they live.

Enchanted Learning: Grasshopper

www.enchantedlearning.com/subjects/insects/orthoptera/Grasshopperprintout.shtml
This page has a printout of grasshopper facts.

INDEX

ABOUT THE AUTHOR

Susan H. Gray is a zoologist who has written many books about animals. She lives in Cabot, Arkansas, with her husband Michael. They have cats that love to chase grasshoppers.